CHEMICAL REACTION

The Essential Guide to Reactions, Acids, Bases, and Metals

UMESH KUMAR

TECHNIC POINT

INTRODUCTION

Chemistry is often viewed as a daunting series of formulas and equations, but it is fundamentally the study of change and the building blocks of our world. This book serves as a focused roadmap through the essential pillars of the subject: **Chemical Reactions**, **Acids, Bases, and Salts**, and **Metallurgy**.

The objective of this guide is to move beyond passive reading. By integrating structured "Concept Capsules" and visual aids, we aim to simplify complex molecular interactions into intuitive lessons. Each chapter is designed to build your confidence, ending with practical challenges that test your grasp of the material and prepare you for advanced academic pursuits.

Whether you are a student or a researcher, this introduction to the core principles of electronics and chemical science provides the clarity needed to master the foundation of the physical sciences.

PREFACE

The goal of this book is to transform the complex world of chemistry into a clear, manageable foundation for every student. Built on the philosophy of **"Subtraction to Multiply,"** this work focuses on stripping away academic clutter to highlight the core principles that truly matter.

As an educator and academic counselor, I have designed this manuscript to bridge the gap between classroom theory and practical exam preparation. By combining unified visual illustrations with "Deep Dive Challenges," this guide encourages students to move beyond rote memorization and toward a genuine understanding of chemical reactions, acids, bases, and metallurgy.

Whether you are preparing for competitive exams or building a foundation for higher education, this book is designed to be your most efficient companion in the pursuit of academic excellence.

Umesh Kumar

CHAPTER 1

CHEMICAL REACTIONS AND EQUATIONS

1.1 Introduction

In our day-to-day life we come across many reactions. Rusting of iron articles, formation of curd from milk and digestion of food in our body all are the examples of chemical reactions. For every reaction there are some common features that are observed. In a chemical reaction one or more following properties must be seen. These are:

a. Change in state

b. Change in color

c. Evolution of gases

d. Change in temperature

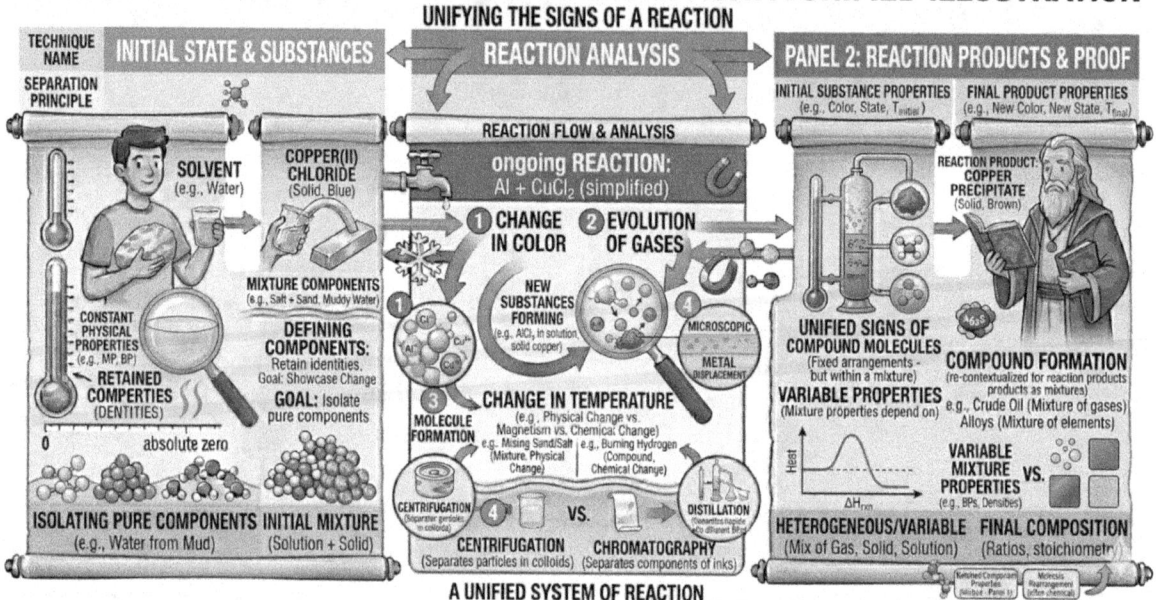

Now, it is easily observable that during the rusting of iron like rime of our bicycle its color changes to brown. it is the change of color. During the formation of curd liquid changes to solid. In the digestion of food almost all changes like change of state, change of color, change of temperature and evolution of gases take place.

1.2 Chemical reaction and equation

Chemical reaction is an actual process while chemical equation is the symbolic representation of chemical reaction.

In a chemical equation the substances which take part are known as reactants and the substances which are formed referred as the product. In an ideal chemical equation, the reactants are kept left side of arrow and added to each other by plus (+) sign and products are kept right side of the arrow.

In a chemical equation the state of the reactants and products are also shown i.e. Solid (s), liquid (l), gas (g), aqueous (aq), precipitate (ppt).

1.3 Balanced chemical equation

In a balanced chemical equation, the number of atoms of each element is equal in reactant sides as well as product side.

If a given reaction is not balanced it is called *skeletal equation.*

1.4 Balancing a chemical equation

Method 1: giving priority to oxygen and hydrogen

For balancing a chemical equation if we give priority to oxygen and then hydrogen it becomes quite easier to balance.

Example 3: balance the following reaction Fe + $O_2 \rightarrow Fe_2O_3$

Here we observe that in LHS there are two atoms of oxygen and 3 atoms of oxygen in RHS. If we multiply LHS by 3 and RHS by 2 the oxygen becomes balanced.

$$Fe + 3O_2 \rightarrow 2Fe_2O_3$$

Now,

We see that oxygen is balanced then for Fe it is 1 in LHS and 4 in RHS therefore multiply by 4 in LHS and reaction becomes balanced.

$$4Fe + 3O_2 \rightarrow 2Fe_2O_3$$

Method 2: an algebraic method

If you are quite interested in mathematics. Here is a handy method for you. Example 4: balance the following equation

Fe + $O_2 \rightarrow Fe_2O_3$

Step1:

Let the coefficient of Fe, O_2 and Fe_2O_3 be a, b and c respectively. Then equation becomes

$$a\, Fe + b\, O_2 \rightarrow c\, Fe_2O_3$$

Step2:

let it is balanced equation then $a = 2c$
and $2b = 3c$
here c has the maximum multipliers so let $c = 1$. On
solving it we get
$A = 2 \quad B = 1.5$
$C = 1$
Then equation becomes $2Fe + 1.5O_2 \rightarrow Fe_2O_3$

Step3:

For getting integral coefficients we can multiply by 2 in LHS and RHS. We get
$4Fe + 3O_2 \rightarrow 2Fe_2O_3$

Note: in many cases step 3 doesn't require.

BALANCING CHEMICAL EQUATIONS: $Fe + O_2 \rightarrow Fe_2O_3$

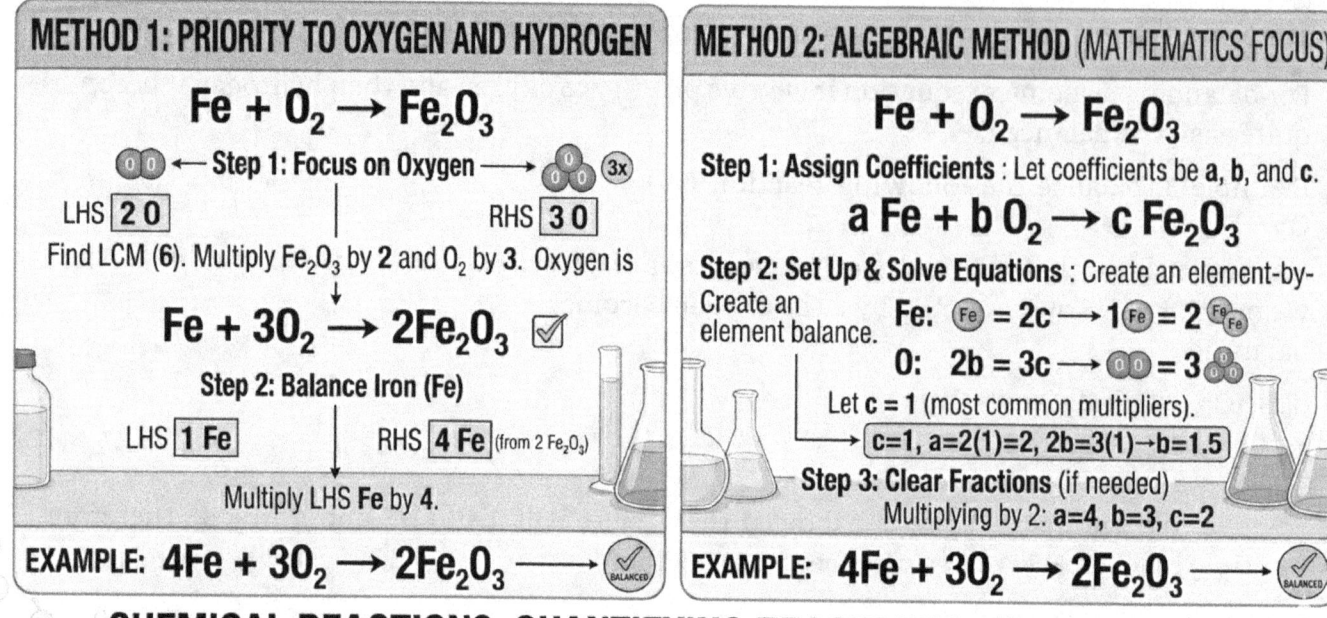

METHOD 1: PRIORITY TO OXYGEN AND HYDROGEN

$$Fe + O_2 \rightarrow Fe_2O_3$$

← Step 1: Focus on Oxygen → (3x)

LHS $\boxed{2\ 0}$ RHS $\boxed{3\ 0}$

Find LCM (6). Multiply Fe_2O_3 by 2 and O_2 by 3. Oxygen is

$$Fe + 3O_2 \rightarrow 2Fe_2O_3 \ \checkmark$$

Step 2: Balance Iron (Fe)

LHS $\boxed{1\ Fe}$ RHS $\boxed{4\ Fe}$ (from 2 Fe_2O_3)

Multiply LHS **Fe** by 4.

EXAMPLE: $4Fe + 3O_2 \rightarrow 2Fe_2O_3$ ⟶ (BALANCED)

METHOD 2: ALGEBRAIC METHOD (MATHEMATICS FOCUS)

$$Fe + O_2 \rightarrow Fe_2O_3$$

Step 1: Assign Coefficients : Let coefficients be **a**, **b**, and **c**.

$$a\ Fe + b\ O_2 \rightarrow c\ Fe_2O_3$$

Step 2: Set Up & Solve Equations : Create an element-by-
Create an
element balance.

Fe: (Fe) $= 2c \rightarrow 1$(Fe)$= 2$ (Fe Fe)

O: $2b = 3c \rightarrow$ (O O) $= 3$

Let **c = 1** (most common multipliers).

$\boxed{c=1,\ a=2(1)=2,\ 2b=3(1) \rightarrow b=1.5}$

Step 3: Clear Fractions (if needed)
Multiplying by 2: **a=4, b=3, c=2**

EXAMPLE: $4Fe + 3O_2 \rightarrow 2Fe_2O_3$ ⟶ (BALANCED)

CHEMICAL REACTIONS: QUANTIFYING REACTANTS AND PRODUCTS

1.5 Types of chemical reactions

1. Combination reaction: when two or more than two reactants combine to form a single product, it is known as combination reaction.

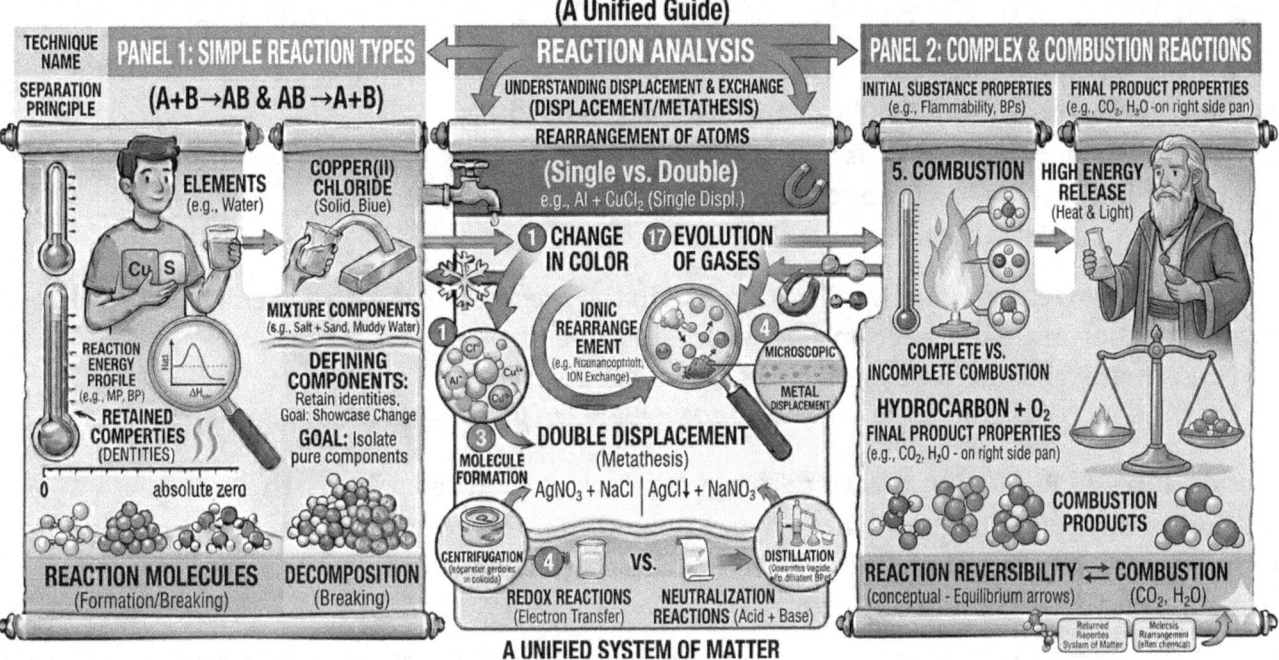

2. Exothermic reaction: the reaction in which heat is released along with the formation of product is known n as exothermic reaction.

Respiration and digestion of food are best example for this.

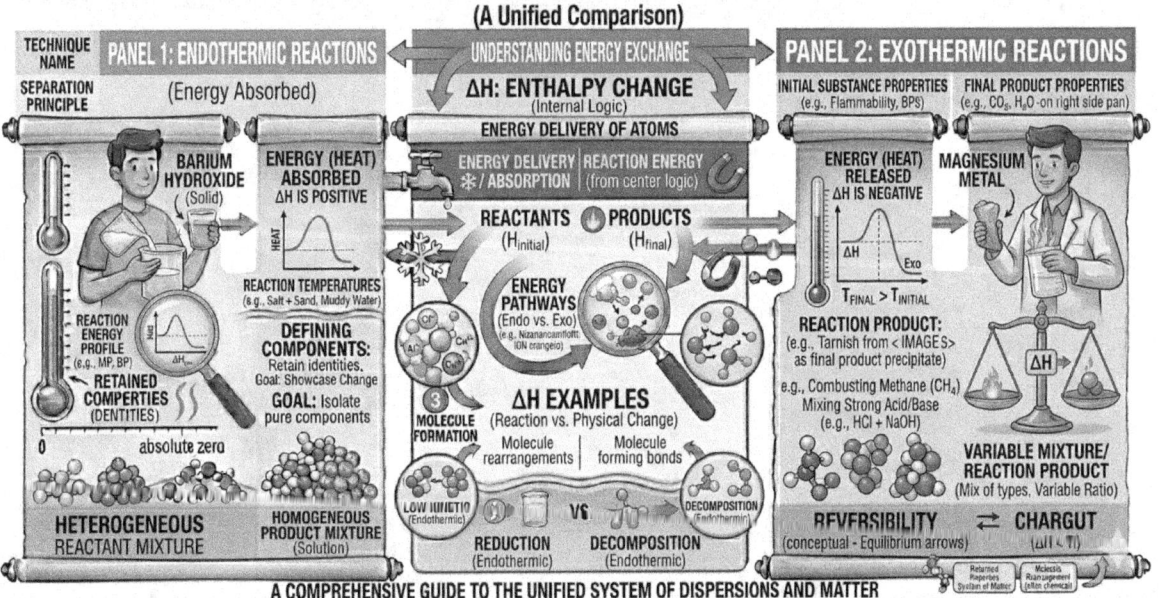

3. **Endothermic reaction:** Endothermic reaction is the reverse process of exothermic. In this reaction the heat is absorbed along the formation of the products.

4. **Decomposition reaction:** in decomposition reaction a single reactant breaks into two or more than two products. This reaction can be carried out using heat or electricity and respectively referred as thermal or electrical decomposition reaction.

5. **Displacement reaction:** this reaction is genuinely reffers to the exchange of ions in when an ionic compound reacts with another higher reactive metal. In actual practice the higher reactive element displaces the lower from its ion.

6. **Double Displacement reaction:** it is similar to the displacement reaction only difference is that in this reaction both reactants are ionic compounds. In this reaction the ions are exchanged.

7. **Precipitation reaction:** in this type of reaction the precipetate is formed which is setteled down to the bottom. This reaction can be also identified in the balanced chemical equation by noticing a downward arrow.

$$Na_2SO_4 \, (aq) + BaCl_2 \, (aq) \rightarrow BaSO_4 \, (ppt) + 2NaCl \, (aq)$$

8. **Oxidation and Reduction reaction:** in a general way oxidation refers to the addition of oxygen or loss of hydrogen and reduction refers to the loss of oxygen or gain of hydrogen. Actually, addition of oxygen removes electron and loss of oxygen refers to the gaining of electron.

In a simple way we can state that in a particular reaction if any atom gaining electron, it is reduced or if it is removing electron it is oxidizing. Oxidation and reduction are the supplementary to each other. Both of these processes are taking place in single reaction, because if there is removal of electrons certainly removed electron is added to another atom participating in the reaction.

1.6 Reactions Affecting Our Life

Corrosion

When any metal comes into the contact of air, water, moisture and acid, it corrodes and this process is known as the corrosion. It is very harmful for our countries economy. In a single year due to corrosion many things made up of iron and other metals corrode and government has to repair it. For example, the government buses, monument's safety bar made up of iron. Our door, the rim of motor vehicles these are also affected.

Corrosion can be minimized using oil, grease on the surface of metallic part, it prevents the air, water, moist and acid to come into the contact of the metal surface. Galvanization is another process to prevent it. In this process a layer of zinc is added to the surface of other metals like iron. As zinc is less prone to corrosion, it is a suitable choice. This done using electrolysis processes.

Rancidity

When oily substances like chips comes into the contact of oxygen get oxidized and their tastes changes. This process is referred as the rancidity. Avoiding these chips packets are flushed with nitrogen gases.

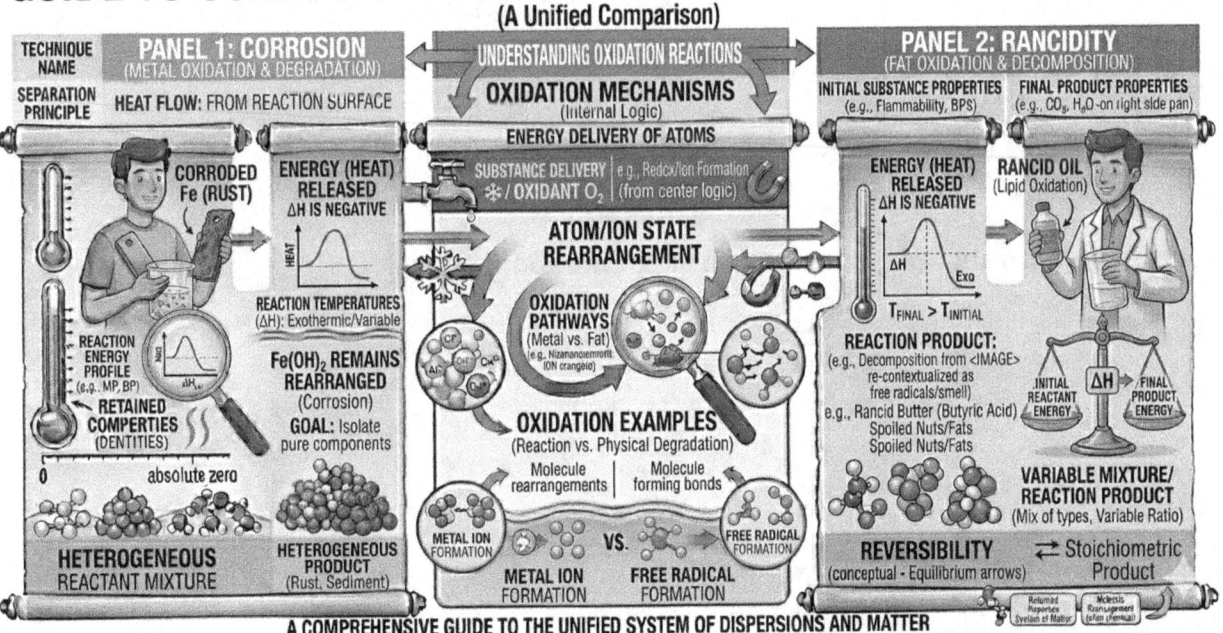

Chapter 1: Deep Dive Challenges

Q1. When Zinc reacts with HCl which gas is released?

(a) H_2
(b) He
(c) O
(d) Cl

Q2. In equation the given reaction, which is reactant?

$H_2 + O_2 \rightarrow H_2O$

(a) H_2
(b) O_2
(c) Both
(d) None

Q3. $Fe + H_2O \rightarrow ? + H_2$

Which compound should be at the place of ? in the above equation?

(a) Fe_2O_3
(b) FeO
(c) Fe_3O_4
(d) $Fe(OH)_3$

Q.4 $3Fe + 4H_2O \rightarrow ? + 4H_2$

Which compound should be at the place of? in the above equation?

(a) Fe_2O_3
(b) FeO
(c) Fe_3O_4
(d) $Fe(OH)_3$

Q5. $C + O_2 \rightarrow CO_2$ is?

(a) Displacement reaction
(b) Endothermic reaction
(c) Exothermic reaction
(d) None

Q6. What is the formula of lime stone?

(a) CaO

(b) $CaCO_2$

(c) $Ca(OH)_2$

(d) $CaCO_3$

Q7. what is the formula of quick lime?

(a) CaO

(b) $CaCo_2$

(c) $Ca(OH)_2$

(d) $CaCO_3$

Q.8 what should be the value of X so that equation becomes balanced?
$2Pb(NO3)2(s) \rightarrow 2PbO(s) + XNO2(g) + O2(g)$

(a) 1

(b) 2

(c) 3

(d) 4

Q.9 $Fe(s) + CuSO4(aq) \rightarrow FeSO4(aq) + Cu(s)$
Which type of reaction is this?

(a) Displacement reaction

(b) combination reaction

(c) decomposition reaction

(d) None

Q.10 which is oxidized in the given reaction?

$2Cu + O_2 \rightarrow CuO$

(a) Cu

(b) O_2

(c) CuO

(d) None

Q.11 Which is essential for corrosion?

(a) air

(b) water

(c) moisture

(d) all of these

Q.12 silver articles become black when exposed to air. In this process which compound is formed?

(a) Silver oxide

(b) Silver nitrate

(c) Silver sulphide

(d) Silver carbonate

Q13. Food kept in air tight jar prevent it from?

(a) oxidation

(b) reduction

(c) corrosion

(d) None

Q14. Chips packets are flushed with?

(a) N_2

(b) O_2

(c) Cl_2

(d) He

Q.15 $2Cu + O_2 \rightarrow 2CuO$
Which type of reaction is this?

(a) Displacement reaction

(b) combination reaction

(c) decomposition reaction

(d) None

Q16. In which type of chemical reaction ions are interchanged?

(a) Displacement reaction

(b) combination reaction

(c) decomposition reaction

(d) double displacement reaction

Q17. Opposite of combination reaction is?

(a) Displacement reaction

(b) combination reaction

(c) decomposition reaction

(d) double displacement reaction

Q18. Oxidation is the gain of?

(a) oxygen

(b) electron

(c) proton

(d) none of these

Q19. Reduction is the gain of?

(a) oxygen

(b) electron

(c) proton

(d) hydrogen

Q20. Precipitates are?

(a) Soluble in water

(b) Insoluble in water

(c) Partially soluble

(d) Float in water

Match the followings:

Q21.

A	B
$AgNO_3$	Combination reaction
$2H_2 + O_2 \rightarrow 2H_2O$	Black
$Fe_2O_3 + Al \rightarrow Al_2O_3 + Fe$	unbalanced reaction
rancidity	nitrogen

Q22.

A	B
oxidation	gain of electron
reduction	loss of electron
corrosion	chips
rancidity	iron

Q23.

A	B
Black and white photography	thermite
reaction Exothermic reaction	$AgBr$
$CuSO_4$	Au

Noble metal	blue

Q24. What is a balanced chemical equation?

Q.25 what is the difference between combination reaction and decomposition reaction?

Q26. what is the difference between exothermic reaction and endothermic reaction?

Q27. what is the difference between displacement reaction and double displacement reaction?

Q28. What are essential conditions for corrosion?

Q29. Why chips packets are flushed with nitrogen gases?

Q30. Write down the reaction which is generally used in black and white photography?

Q31. What is rancidity? Explain with examples.

Q32. Why we apply paints on metal articles?

Q33. A shiny brown colored element 'X' on heating in air becomes black in color. Name the element 'X' and the black-colored compound formed.

Long answer type questions

Q34. Translate the following statements into chemical equations and then balance them.

(a) Hydrogen gas combines with nitrogen to form ammonia.

(b) Hydrogen sulphide gas burns in air to give water and Sulphur dioxide.

(c) Barium chloride reacts with aluminum sulphate to give aluminum chloride and a precipitate of barium sulphate.

(d) Potassium metal reacts with water to give potassium hydroxide and hydrogen gas.

Q35. Balance the following chemical equations.

(a) $HNO_3 + Ca(OH)_2 \rightarrow Ca(NO_3)_2 + H_2O$

(b) $NaOH + H_2SO_4 \rightarrow Na_2SO_4 + H_2O$

(c) $NaCl + AgNO_3 \rightarrow AgCl + NaNO_3$

(d) $BaCl_2 + H_2SO_4 \rightarrow BaSO_4 + HCl$

Q36. Write the balanced chemical equations for the following reactions.

(a) Calcium hydroxide + Carbon dioxide → Calcium carbonate + Water

(b) Zinc + Silver nitrate → Zinc nitrate + Silver

(c) Aluminum + Copper chloride → Aluminum chloride + Copper

(d) Barium chloride + Potassium sulphate → Barium sulphate + Potassium chloride

Q37. Write the balanced chemical equation for the following and identify the type of reaction in each case.

(a) Potassium bromide(aq) + Barium iodide(aq) → Potassium iodide(aq) + Barium bromide(s)

(b) Zinc carbonate(s) → Zinc oxide(s) + Carbon dioxide(g)

(c) Hydrogen(g) + Chlorine(g) → Hydrogen chloride(g)

(d) Magnesium(s) + Hydrochloric acid(aq) → Magnesium chloride(aq) + Hydro-
    gen(g)

Solution

1.A 2.C 3.A 4.C 5.C 6. A 7.C 8. B 9.A 10. A 11.D 12.
B 13.A 14. A 15.B 16. D 17.C 18. A

19.B 20. B

Short Answer Hints

- Q24: Think about the Law of Conservation of Mass. Are the number of atoms for each element the same on the left and right?
- Q25: One is about "joining" $(A + B \rightarrow AB)$, while the other is about "breaking down" $(AB \rightarrow A + B)$.
- Q26: Focus on the flow of heat. Does the test tube feel hot (releasing energy) or cold (absorbing energy)?
- Q27: In one, a "stronger" element kicks out a "weaker" one. In the other, two compounds simply exchange partners.
- Q28: Metals need two "partners" from the environment to rust: one is a liquid and one is a gas found in the air.
- Q29: Oxygen causes food to spoil. Nitrogen is inert (unreactive) and acts as a protective cushion.
- Q30: Look for the decomposition of silver halides (Silver Chloride or Silver Bromide) when exposed to sunlight.
- Q31: This involves the oxidation of fats and oils in food, leading to a bad smell and taste. Think of old, "stale" butter.
- Q32: Paint acts as a barrier. If air and moisture can't touch the metal, what process is prevented?
- Q33: Element 'X' is a common metal used in electrical wires. When it reacts with oxygen (O_2), it forms a black oxide.

Long Answer Hints

Q34: Translation and Balancing

- a) $H_2 + N_2 \rightarrow NH_3$. (Hint: You need 2 Nitrogen atoms on the right).
- (b) $H_2S + O_2 \rightarrow H_2O + SO_2$. (Hint: Oxygen appears in two places on the right; count carefully!)
- (c) $BaCl_2 + Al_2(SO_4)_3 \rightarrow AlCl_3 + BaSO_4$. (Hint: Balance the Sulfate (SO_4) groups as a single unit).
- (d) $K + H_2O \rightarrow KOH + H_2$. (Hint: Use a coefficient of 2 for K, H_2O, and KOH).

Q35: Balancing Equations

- (a) Look at the Nitrate (NO_3) group; there are two on the right.
- (b) Start by balancing the Sodium (Na) atoms first.
- (c) Check the atoms carefully—is this one already balanced?

- (d) Balance the Chlorine (Cl) and Hydrogen (H) by adding a single coefficient on the right.

Q36: Writing Balanced Equations

- (a) $Ca(OH)_2 + CO_2 \rightarrow CaCO_3 + H_2O$.
- (b) Zinc is more reactive than Silver; it will take the Nitrate group.
- (c) Aluminum has a valency of 3, while Copper is usually 2. This affects the subscripts in $AlCl_3$ and $CuCl_2$.
- (d) This is a partner-exchange reaction. Ensure the formulas for Barium Sulfate and Potassium Chloride are correct.

Q37: Balancing and Identification

- (a) Type: Double Displacement. (Hint: Ions are being exchanged).
- (b) Type: Decomposition. (Hint: A single reactant is breaking into two products).
- (c) Type: Combination/Synthesis. (Hint: Two elements are forming one compound).
- (d) Type: Displacement. (Hint: A metal is reacting with an acid to release a gas).

CHAPTER 2

ACIDS, BASES AND SALTS

2.1 Formation of acids

When nonmetals react with oxygen it produces nonmetallic oxide.

$$S(s) + O_2(g) \rightarrow SO_2(g)$$

$$P_4(s) + 5O_2(g) \rightarrow P_4O_{10}(s)$$

$$2H_2(g) + O_2(g) \rightarrow 2H_2O(l)$$

When nonmetallic oxide reacts with water acid is formed.

$$SO_3(g) + H_2O(l) \rightarrow H_2SO_4(aq)$$

$$P_4O_{10}(s) + 6H_2O(l) \rightarrow 4H_3PO_4(aq)$$

$$4NO_2(g) + 2H_2O(l) + O_2(g) \rightarrow 4HNO_3(aq)$$

2.2 Formation of bases

When metals react with oxygen it produces metallic oxide.

$$4Fe + 3O_2 \rightarrow 2Fe_2O_3$$
$$2Cu + O_2 \rightarrow 2CuO$$
$$4Na + O_2 \rightarrow 2Na_2O$$
$$4Al + 3O_2 \rightarrow 2Al_2O_3$$

When metallics oxide reacts with water base is formed.

$$CaO(s) + H_2O(l) \rightarrow Ca(OH)_2(aq)$$
$$K_2O(s) + H_2O(l) \rightarrow 2KOH(aq)$$
$$MgO(s) + H_2O(l) \rightarrow Mg(OH)_2(aq)$$

In a general way we can understand the formation of base by a very simple activity.

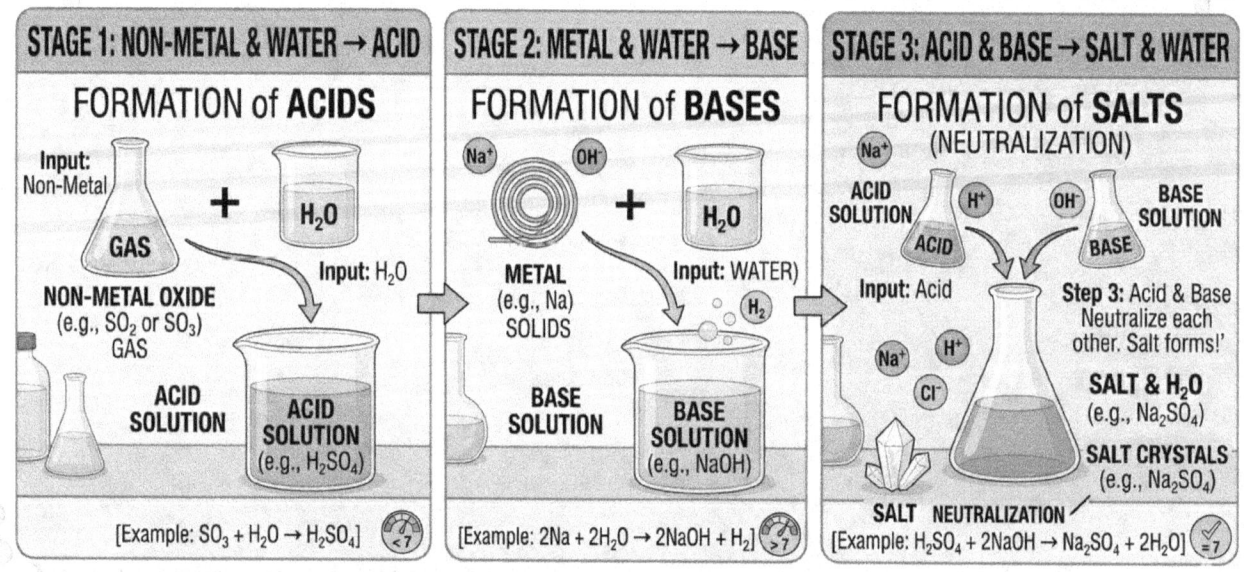

Activity 1:

Step 1: take a magnesium ribbon and clean its upper surface by sand paper.

Step2: burn it with the help of a pair of tongs and a spirit lamp.

Step 3: collect its ash and dissolve in a glass of water.

Now you have a base in your glass.

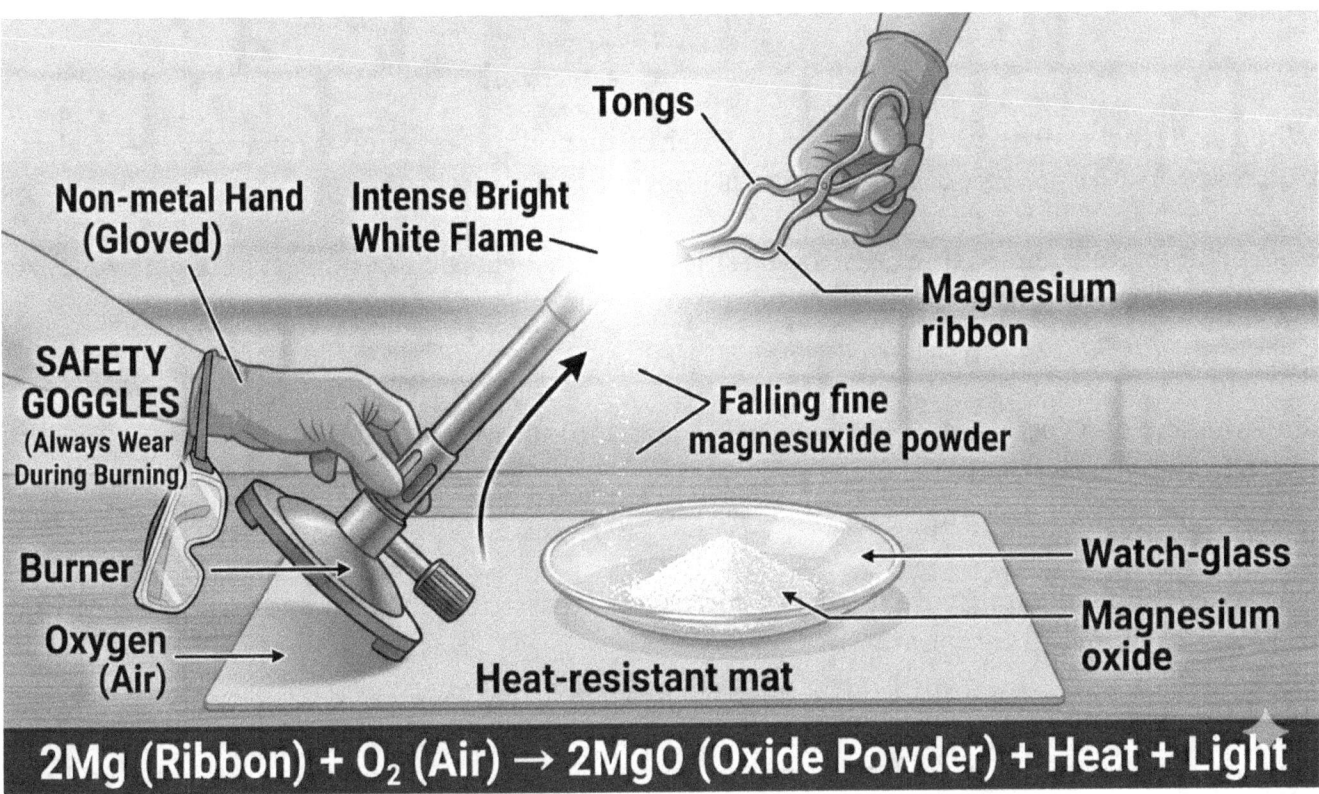

2Mg (Ribbon) + O$_2$ (Air) → 2MgO (Oxide Powder) + Heat + Light

2.3 Formation of salts

When acid reacts with base, both neutralize each other and this reaction is referred as neutralization reaction. The product of this reaction is known as salt.

Example:

Hydrochloric Acid and Sodium HydroxideThis is the most classic example. When hydrochloric acid (strong acid) reacts with sodium hydroxide (strong base), they produce common table salt and water.

Reaction: $HCl(aq) + NaOH(aq) \rightarrow NaCl(aq) + H_2O(l)_2$.

Sulfuric Acid and Potassium HydroxideIn this reaction, sulfuric acid reacts with potassium hydroxide to form potassium sulfate (a salt) and water.

Reaction: $H_2SO_4(aq) + 2KOH(aq) \rightarrow K_2SO_4(aq) + 2H_2O(l)_3$.

Nitric Acid and Calcium HydroxideWhen nitric acid reacts with calcium hydroxide (slaked lime), it produces calcium nitrate and water.

Reaction: $2HNO_3(aq) + Ca(OH)_2(aq) \rightarrow Ca(NO_3)_2(aq) + 2H_2O(l)$

Now we have a general idea of formation of acids, bases and salt. In the next section we will discuss some physical and chemical properties of acid, base and salt.

2.4 Detection of acids and bases in general life

Generally, acids are sour in taste while bases are bitter in nature. Acid turns blue litmus paper to red while base changes red to blue. When acids and bases are in solution or aqueous form we can check whether they are acids or bases using many natural flowers and petals. These flowers or petals are referred as natural indicator. Some of these are red cabbage, turmeric, colored petals of flowers such as hydrangea, petunia and geranium. There are many synthesized indicators. These are shown in the following table.

Indicator	pH range (color change)	color of acid form	color of conjugate base form
methyl orange	3,1 - 4,4	orange	yellow
methyl red	4,2 - 6,2	red	yellow
bromothymol blue	6,0 - 7,8	yellow	blue
phenolphthalein	8,3 - 10,0	colourless	pink
alizarin yellow	10,1 - 12,1	yellow	red

2.5 Chemical Properties of Acids and Bases

Reaction with metals

When acids or bases react with metal salt and hydrogen gas is produced.

$$NaOH + Zn \rightarrow Na_2ZNO_2 + H_2$$

Reaction with metal carbonates and hydrogen carbonates

On reaction with metal carbonate or metal hydrogen carbonate acids and bases produces salt and water with carbon di oxide gas.

1. Reaction with Metal Carbonates.

The general equation is:
Acid + Metal Carbonate→Salt + Water + Carbon Dioxide Example: Hydrochloric Acid and Sodium CarbonateWhen dilute hydrochloric acid is added to sodium carbonate, sodium chloride, water, and carbon dioxide are formed. Reaction: $Na_2CO_3(s) + 2HCl(aq) \rightarrow 2NaCl(aq) + H_2O(l) + CO_2(g)$

2. Reaction with Metal Hydrogen CarbonatesThe general equation is:
Acid + Metal Hydrogen Carbonate→Salt + Water + Carbon Dioxide Example: Hydrochloric Acid and Sodium Hydrogen CarbonateWhen hydrochloric acid reacts with sodium hydrogen carbonate (baking soda), the products are the same as above. Reaction: $NaHCO_3(s) + HCl(aq) \rightarrow NaCl(aq) + H_2O(l) + CO_2(g)$

How to Test for Carbon Dioxide (CO_2)

To confirm that the gas produced is carbon dioxide, it is passed through lime water (calcium hydroxide). The lime water turns milky due to the formation of a white precipitate of calcium carbonate.
Test Reaction: $Ca(OH)_2(aq) + CO_2(g) \rightarrow CaCO_3(s)\downarrow + H_2O(l)$

Metallic oxides with acids

As metallic oxide is basic in nature when it reacts with any acid salt is formed with the release of hydrogen gas.

Nonmetallic oxides with base

As nonmetallic oxide is acidic in nature when it reacts with any base salt is formed with the release of hydrogen gas.

Acids and bases in aqueous state

When acid or bases are diluted in water, they release H^+ or H^- ions respectively. These ions are necessary to give authentic pH test or any indicator test. That's why all test relevant to acids and bases are taken in aqueous state only.

The following table will recapitulate the properties of acids and bases:

RECAPITULATING THE PROPERTIES OF ACIDS AND BASES: KEY CHEMICAL REACTIONS

REACTION TYPE		REACTANTS	PRODUCTS	SUBSTANCES FORMULA
Acid-Metal Reaction Forms Hydrogen Gas		Acid + Metal	Salt + H_2	e.g., $HCl + Mg \rightarrow MgCl_2 + H_2$
Neutralization (Hydroxide)		Acid + Metal Hydroxide	Salt + H_2O	e.g., $HCl + NaOH \rightarrow NaCl + H_2O$
Neutralization (Oxide)		Acid + Metal Oxide	Salt + H_2O	e.g., $H_2SO_4 + CuO \rightarrow CuSO_4 + H_2O$
Acid-Carbonate Reaction (Forms CO_2)		Acid + Metal Carbonate	Salt + H_2O + CO_2	e.g., $HCl + CaCO_3 \rightarrow CaCl_2 + H_2O + CO_2$
Acid-Bicarbonate Reaction (Forms CO_2)		Acid + Metal Hydrogen Carbonate	Salt + H_2O + CO_2	e.g., $HCl + NaHCO_3 \rightarrow NaCl + H_2O + CO_2$
Neutralization (Acidic Oxide)		Acidic Oxide + Base	Salt + H_2O	e.g., $CO_2 + Ca(OH)_2 \rightarrow CaCO_3 + H_2O$

CONSOLIDATED REACTION SUMMARY

2.6 pH Scales a Strong Tool

pH scale is a scale designed to test the acidic or basic character of any substance. In this scale there are 0 to 14 readings. In this reading 7 refers to the neutral substance while less is acidic and more is basic in nature.

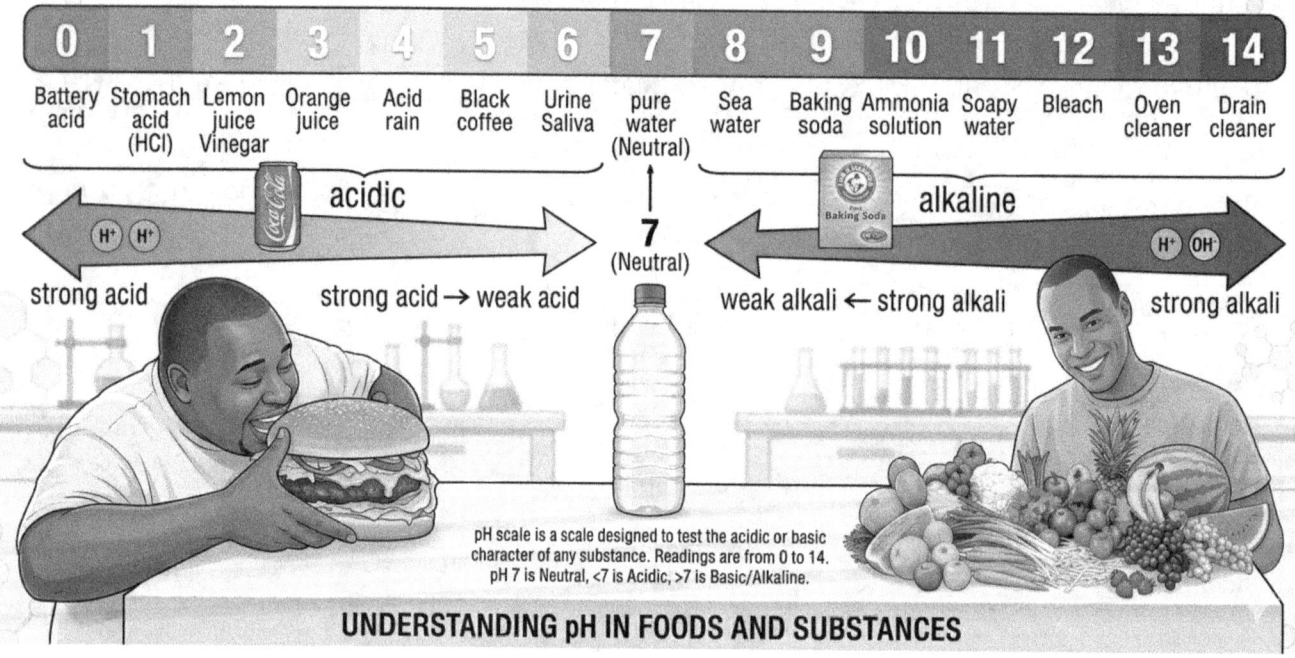

2.7 Some important salts

Bleaching powder

Formation

Chlorine gas is used in the formation of bleaching powder. Bleaching powder is produced by the action of chlorine on dry slaked lime [$Ca(OH)_2$].

$$Ca(OH)_2 + Cl_2 \rightarrow CaOCl_2 + H_2O$$

Uses

(i)   for bleaching cotton and linen in the textile industry

(ii)   for bleaching wood pulp in paper factories

(iii)   for bleaching washed clothes in laundry

(iv)   As an oxidizing agent in many chemical industries

(v)   For disinfecting drinking water to make it free of germs

Baking soda Formation

In the formation of baking soda i.e. sodium hydrogen carbonates the basic raw material is sodium chloride.

$$NaCl + H_2O + CO_2 + NH_3 \rightarrow NH_4Cl + NaHCO_3$$

During cooking

$$2NaHCO_3 \rightarrow Na_2CO_3 + H_2O + CO_2$$

Sodium hydrogen carbonate has got various uses in the household.

Uses of sodium hydrogen carbonate ($NaHCO_3$)

(i) For making baking powder, this is a mixture of baking soda (sodium hydrogen carbonate) and a mild edible acid such as tartaric acid. When baking powder is heated or mixed in water, the following reaction takes place –

$$NaHCO_3 + H^+ \rightarrow CO_2 + H_2O + \text{Sodium salt of acid}$$

Carbon dioxide produced during the reaction causes bread or cake to rise making them soft and spongy.

(ii) Sodium hydrogen carbonate is also an ingredient in antacids. Being alkaline, it neutralizes excess acid in the stomach and provides relief.

(iii) It is also used in soda-acid fire extinguishers.

Washing soda Formation

When sodium hydrogen carbonate is heated it gives sodium carbonate. On recrystallisation i.e. adding 10 molecules of water to sodium carbonate gives washing soda.

$$Na_2CO_3 + 10.H_2O \rightarrow Na_2CO_3.10H_2O$$

Uses

(i) Sodium carbonate (washing soda) is used in glass, soap and paper industries.

(ii) It is used in the manufacture of sodium compounds such as borax.

(iii) Sodium carbonate can be used as a cleaning agent for domestic purposes.

(iv) It is used for removing permanent hardness of water.

Plaster of Paris

Plaster of Paris (PoP) is chemically known as Calcium Sulfate Hemihydrate. It is prepared by heating gypsum ($CaSO_4 \cdot 2H_2O$) at a very specific temperature of 373 K (100°C). At this temperature, gypsum loses three-quarters of its water of crystallization to become PoP. When you add water back to Plaster of Paris, it undergoes a hydration reaction to re-form gypsum, which sets into a very hard, rigid mass. The Chemical Reaction The equation for this transformation is:

$$CaSO_4 \cdot \frac{1}{2}H_2O + 1\frac{1}{2}H_2O \rightarrow CaSO_4 \cdot 2H_2O$$

Plaster of Paris + Water $\rightarrow$ Gypsum (Hard Mass)

Key Points about the Reaction: Water of Crystallization: In Plaster of Paris, two formula units of $CaSO_4$ share one molecule of water. This is why it is written as $CaSO_4 \cdot \frac{1}{2}H_2O$.

Setting Time: Once mixed with water, the paste remains workable for a short period before it begins to "set" or harden into the solid gypsum structure.

Storage: Because it reacts so readily with moisture in the air to become hard, Plaster of Paris must be stored in moisture-proof containers.

Common Uses

Because of its ability to set into a hard, molded shape, it is used extensively in various fields:

Medical: Doctors use it as a bandage for supporting fractured bones in the right position.

Construction & Decor: It is used for making smooth surfaces on walls and creating decorative ceiling designs (false ceilings).

Art: It is a popular material for making toys, statues, and casts for dental molds.

Laboratory: It can be used to seal air gaps in laboratory apparatus to make them airtight.

It is used for making toys, materials for decoration and for making surfaces smooth.

FORMATION AND USES OF COMMON SALTS

BLEACHING POWDER
(Calcium Hypochlorite)

$CaOCl_2$

FORMATION

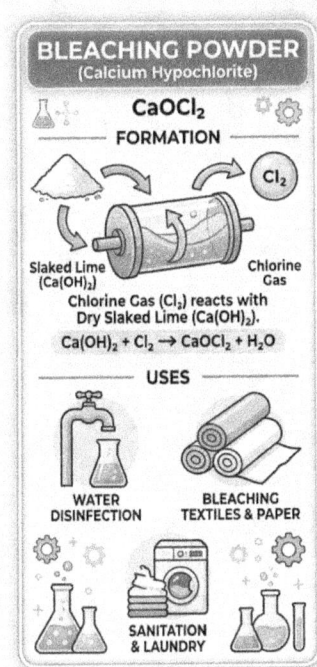

Slaked Lime $(Ca(OH)_2)$ Chlorine Gas

Chlorine Gas (Cl_2) reacts with Dry Slaked Lime $(Ca(OH)_2)$.

$Ca(OH)_2 + Cl_2 \rightarrow CaOCl_2 + H_2O$

USES

- WATER DISINFECTION
- BLEACHING TEXTILES & PAPER
- SANITATION & LAUNDRY

BAKING SODA
(Sodium Bicarbonate)

$NaHCO_3$

FORMATION

Solvay Process:

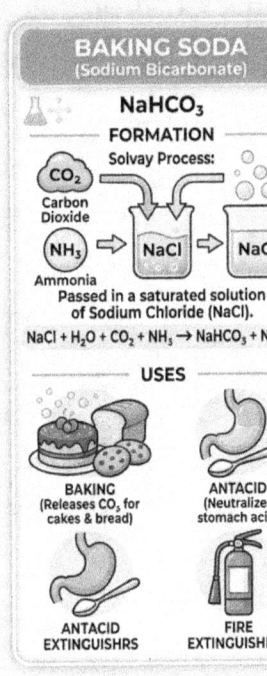

CO_2 Carbon Dioxide

NH_3 Ammonia

Passed in a saturated solution of Sodium Chloride (NaCl).

$NaCl + H_2O + CO_2 + NH_3 \rightarrow NaHCO_3 + NH_4Cl$

USES

- BAKING (Releases CO_2 for cakes & bread)
- ANTACID (Neutralizes stomach acid)
- ANTACID EXTINGUISHRS
- FIRE EXTINGUISHERS

WASHING SODA
(Sodium Carbonate)

$Na_2CO_3 \cdot 10H_2O$

FORMATION

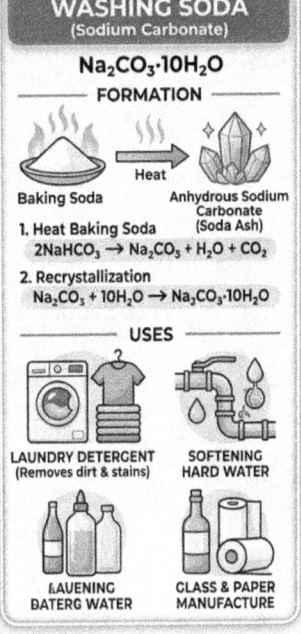

Baking Soda   Heat   Anhydrous Sodium Carbonate (Soda Ash)

1. Heat Baking Soda
$2NaHCO_3 \rightarrow Na_2CO_3 + H_2O + CO_2$

2. Recrystallization
$Na_2CO_3 + 10H_2O \rightarrow Na_2CO_3 \cdot 10H_2O$

USES

- LAUNDRY DETERGENT (Removes dirt & stains)
- SOFTENING HARD WATER
- LAUENING DATERG WATER
- GLASS & PAPER MANUFACTURE

PLASTER OF PARIS
(Calcium Sulphate Hemihydrate)

$CaSO_4 \cdot \frac{1}{2}H_2O$

FORMATION

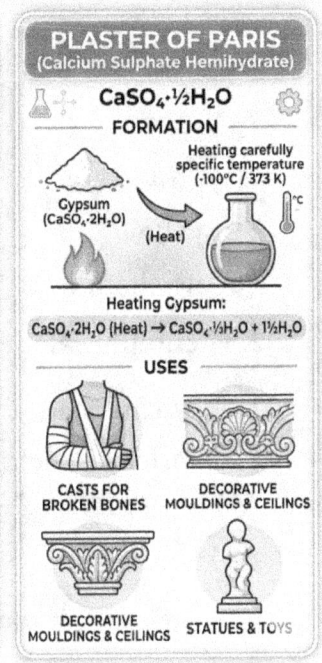

Heating carefully specific temperature (·100°C / 373 K)

Gypsum $(CaSO_4 \cdot 2H_2O)$ (Heat)

Heating Gypsum:
$CaSO_4 \cdot 2H_2O \text{ (Heat)} \rightarrow CaSO_4 \cdot \frac{1}{2}H_2O + 1\frac{1}{2}H_2O$

USES

- CASTS FOR BROKEN BONES
- DECORATIVE MOULDINGS & CEILINGS
- DECORATIVE MOULDINGS & CEILINGS
- STATUES & TOYS

Chapter 2: Deep Dive Challenges

Q1. Acid changes blue litmus to?

(a) red
(b) pink
(c) yellow
(d) no change

Q2. Phenolphthalein and methyl oranges are

(a) synthetic indicator
(b) natural indictor
(c) both
(d) none

Q3. In olfactory indicators?

(a) Colour changes
(b) Taste changes
(c) Odour changes
(d) none

Q4. Acid on reaction with metal produces?

(a) H_2
(b) O_2
(c) N_2
(d) None

Q5. Acid on reaction with metal hydrogen carbonate produces?

(a) H_2
(b) CO_2
(c) N_2
(d) None

Q6. In aqueous solution acid produces which type of ions?

(a) H^+
(b) OH^-
(c) N_2
(d) None

Q7. The process of dissolving an acid or a base in water is highly?

(a) endothermic
(b) exothermic
(c) cold
(d) None

Q8. pH scale measure which ion in a solution?

(a) H^+
(b) OH^-
(c) Cl^-
(d) None

Q9. A solution has pH value 7. It is likely to be

(a) acid
(b) base
(c) salt
(d) none of these

Q10. Our body works within pH range?

(a) 2-4
(b) 12-14
(c) 6-10
(d) 7-7.8

Q11. Milk of magnesia is?

(a) acid
(b) weak acid
(c) antacid
(d) None

Q12. Acetic acid is present in?

(a) Apple
(b) guava
(c) vinegar
(d) None

Q13. Oxalic acid is present in?

(a) potato

(b) tomato

(c) onion

(d) none

Q14. pH value of a salt is?

(a) 6

(b) 7

(c) 8

(d) 9

Q15. Formula for bleaching powder is?

(a) $CaOCl_2$

(b) $CaOCl$

(c) $CaCl_3$

(d) None

Q16. Na_2CO_3 is?

(a) Washing soda

(b) Sodium carbonate

(c) Sodium hydrogen carbonate

(d) None

Q17. Which is used for removing permanent hardness of water?

(a) $CaSO_4.1/2H_2O$

(b) $CaSO_4.2H_2O$

(c) $CaSO_4$

(d) Washing soda

Q18. A solution turns blue litmus red its pH is likely to be?

(a) 1

(b) 7

(c) 8

(d) 9

Q19. A solution reacts with crushed egg-shells to give a gas that turns lime-water milky. The solution contains

(a) NaCl

(b) HCl

(c) LiCl

(d) KCl

Q20. Which one of the following types of medicines is used for treating indigestion?

(a) Antibiotic

(b) Analgesic

(c) Antacid

(d) Antiseptic

Match the following:

Q21.

A	B
Acetic acid	Nettle sting
Citric acid	Curd
Lactic acid	Vinegar
Methanoic acid	Orange

Q22.

A	B
Bleaching powder	$CaSO_4.2H_2O$
Baking soda	$CaSO_4.1/2H_2O$
Washing soda	$CaOCl_2$
Plasterof paris	$Na_2CO_3.10H_2O$
Gypsum	$NaHCO_3$

Q23.

A	B
Acid	Red cabbage
Base	Salt
pH	H^+
Natural indicator	OH^-

Short Answer Questions

Q24. Write down physical differences between acid and bases.

Q25. Write down chemical differences between acid and bases.

Q26. What are common properties between acid and base?

Q27. What is salt and how it is formed?

Q28. What is anta-acids? How they help us in digestion?

Q29. Write down the uses of baking soda?

Q30. How does plaster of Paris is formed?

Q31. Why should curd and sour substances not be kept in brass and copper vessels?

Q32. Why does dry HCl gas not change the color of the dry litmus paper?

Long answer type questions:

Q33. Suppose you are sent to a planet where total soil is acidic in nature. It is also not suitable for agriculture purpose. You have to grow plant there. What procedure you can use to solve this problem?

Q34. Write down the formation and uses of following compounds:

(a) Washing soda

(b) Baking soda

(c) Bleaching powder

(d) Plaster of paris

Q35. Write down five chemical properties of acids. Explai with chemical equation in each case.

Q36. Write down five chemical properties of bases. Explai with chemical equation in each case.

Solution

1.a 2.a 3.c 4.a 5.b 6 a 7.b 8.a 9.c 10.d 11.c 12.c 13.b 14.b 15.a 16.b 17.d 18.a 19.b 20.c

---

Short Answer Hints
- Q24 (Physical): Think about taste (sour vs. bitter) and touch (one can be corrosive, the other feels soapy).
- Q25 (Chemical): Focus on the specific ions they release in water (H^+ vs. OH^-) and how they change the color of Litmus paper.
- Q26 (Commonalities): Think about their ability to conduct electricity in solution and their shared goal in a neutralization reaction.
- Q27 (Salt): It's the "offspring" of a chemical marriage. What happens when an acid and a base cancel each other out?
- Q28 (Antacids): These are mild bases. If your stomach has too much "fire" (acid), how does a base help cool it down?

- Q29 (Baking Soda): Think of fluffy cakes (releasing CO2), fire extinguishers, and its use as a mild antiseptic.
- Q30 (P.O.P.): It comes from a mineral called Gypsum. What happens when you carefully heat Gypsum at exactly 373 K?
- Q31 (Vessels): Sour substances contain acids. What happens when an acid reacts with a metal like Copper? Is the resulting product safe to eat?
- Q32 (Dry HCl): Acids only show their "true colors" (ionize) in the presence of water. If everything is dry, can the H^+ ions move?

Long Answer Hints

Q33: The Acidic Planet

- The Problem: The soil pH is too low for plants.
- The Solution: Think about Neutralization. What "household" or industrial basic substances (like Lime or Chalk) could you add to the soil to raise the pH?

Q34: Formation and Uses

- (a) Washing Soda: Look up the Solvay Process and the recrystallization of Sodium Carbonate. (Use: Laundry/Glass making).
- (b) Baking Soda: Reaction of Brine with CO2 and NH3. (Use: Antacid/Baking).
- (c) Bleaching Powder: Action of Chlorine gas on Dry Slaked Lime. (Use: Disinfecting water).
- (d) Plaster of Paris: Half-hydration of Calcium Sulphate. (Use: Fixing fractured bones).

Q35: Chemical Properties of Acids

Think about what happens when an acid meets:

1. Metal: (Releases Hydrogen gas).
2. Metal Carbonate: (Releases Carbon Dioxide).
3. Base: (Forms Salt and Water).
4. Metal Oxide: (Basic in nature, so think neutralization).
5. Water: (Heat is released; always add acid to water, not water to acid!).

Q36: Chemical Properties of Bases

Think about what happens when a base meets:

1. Certain Metals: (Like Zinc or Aluminum—releases Hydrogen).
2. Acid: (Neutralization).
3. Non-Metal Oxide: (Non-metal oxides are acidic, so what is the result?).
4. Ammonium Salts: (Often releases Ammonia gas).
5. Water: (Dissolving to form Alkalis).

CHAPTER 3

Metals and Non-Metals

3.1 Introduction

Physical properties of metals and non-metals

3.1 Hardness

Hardness refers to strength of a material. Generally, metals are hard and non-metals are soft and brittle. There are some exceptions, lead is a brittle metal. Some metals are so soft that they can be cut by knife such as sodium, potassium, magnesium.

3.2 Malleability

Malleability is the property of being beaten in the form of a thin sheet. Metals can be beaten in the thin sheets. This property helps us to use metals sheets, containers etc.

3.3 DUCTILITY

Metals can be drawn into the wires this property is known as ductility. Gold is most ductile element that's why ornaments are not form from the pure gold.

3.4 Conduction of heat

The materials which allow heat to pass through them are known as the conductor of heat. Metals are good conductor of heat and nonmetals are bad.

3.5 Conduction of electricity

The materials which allow electricity to pass through them are known as the conductor of electricity. Metals are good conductors and nonmetals are insulator. Graphite is a nonmetal but a good conductor of electricity.

3.6 Sonority

Sonority refers to the sounds produced when a metal is beaten. Due to this property copper or irons are used as a bell.

3.7 Metallic Luster

Metals have a shining surface while nonmetals have not. This property is referred as metallic luster. Iodine is exception; it is lustrous but nonmetal.

COMPARATIVE PHYSICAL PROPERTIES: METALS VS. NON-METALS

METALS		NON-METALS	
iron bar + anvil + strong hammer	GENERALLY HARD, DENSE (Except Na, K) — **HARDNESS** — GENERALLY SOFT, LOW DENSITY (Diamond is hard)	carbon, sulfur, diamond (exception)	
	CAN BE HAMMERED INTO SHEETS — **MALLEABILITY** — BRITTLE, SHATTER WHEN STRUCK	coal + sulfur	
	CAN BE DRAWN INTO WIRES — **DUCTILITY** — NOT DUCTILE		
	GOOD CONDUCTORS OF HEAT — **HEAT CONDUCTION** — POOR CONDUCTORS OF HEAT	or	
or	GOOD CONDUCTORS OF ELECTRICITY (Free electrons) — **ELECTRICITY CONDUCTION** — POOR CONDUCTORS (Graphite is an exception)	or	
or	SONOROUS (Ring when struck) — **SONORITY** — NOT SONOROUS		
	SHINY AND LUSTROUS — **METALLIC LUSTRE** — GENERALLY DULL		

*Exceptions apply in some cases

3.8 Chemical properties of metals

In the previous lesson we have seen the reaction of metals and non-metals with acids, bases, carbonates, hydrogen carbonates, and oxides. In this section we will see what happens when metals and a non-metal react.

3.9 Reaction with solution of other salts

When metals react with solution of the other salts displacement reaction occurs. About we have already discussed in the chemical reaction part. The following reactivity series will help in finding the more reactive and less reactive elements.

3.10 The reactivity series

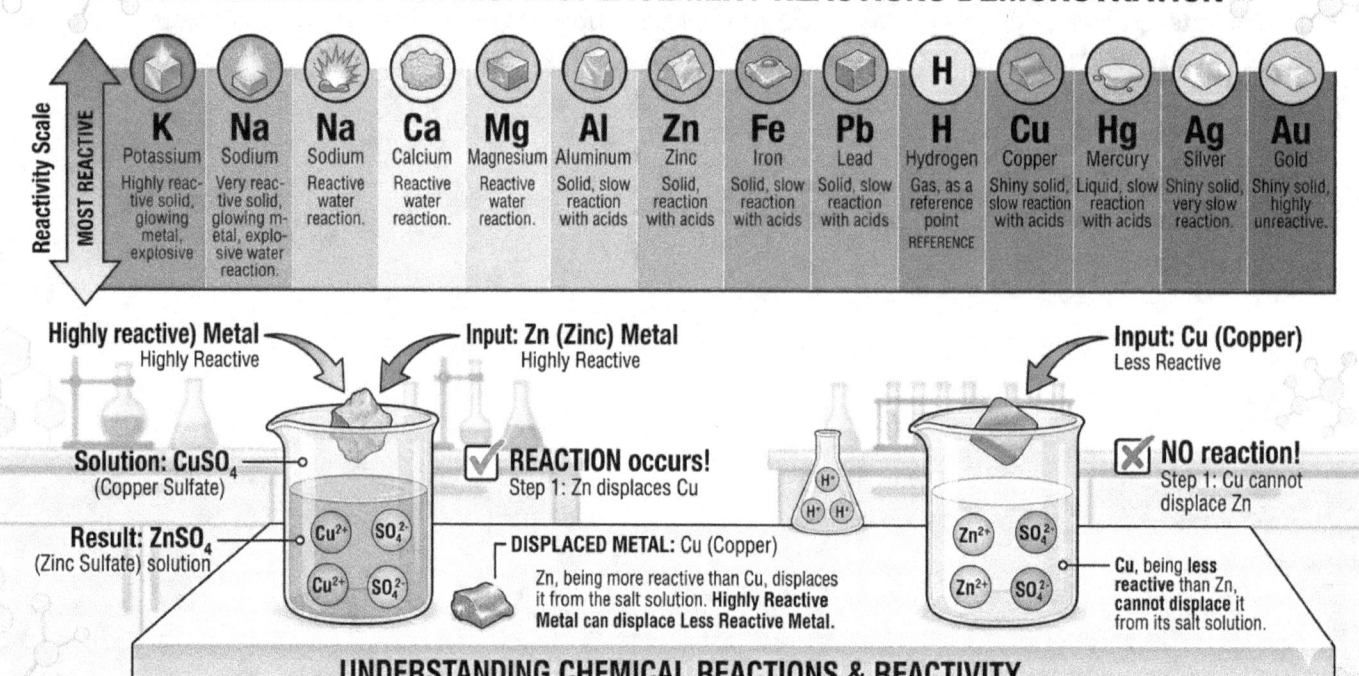

3.11 Reaction with non-metals

In the first part of this series, in chapter 0 (some basic concepts) we have studied that when metals and nonmetals come into contact electrons are transferred from metal to nonmetal and anions and cations are formed. These are attracted due to opposite polarity and a strong ionic bond is formed. These compounds are referred as ionic compounds.

3.12 Ionic compounds and its properties

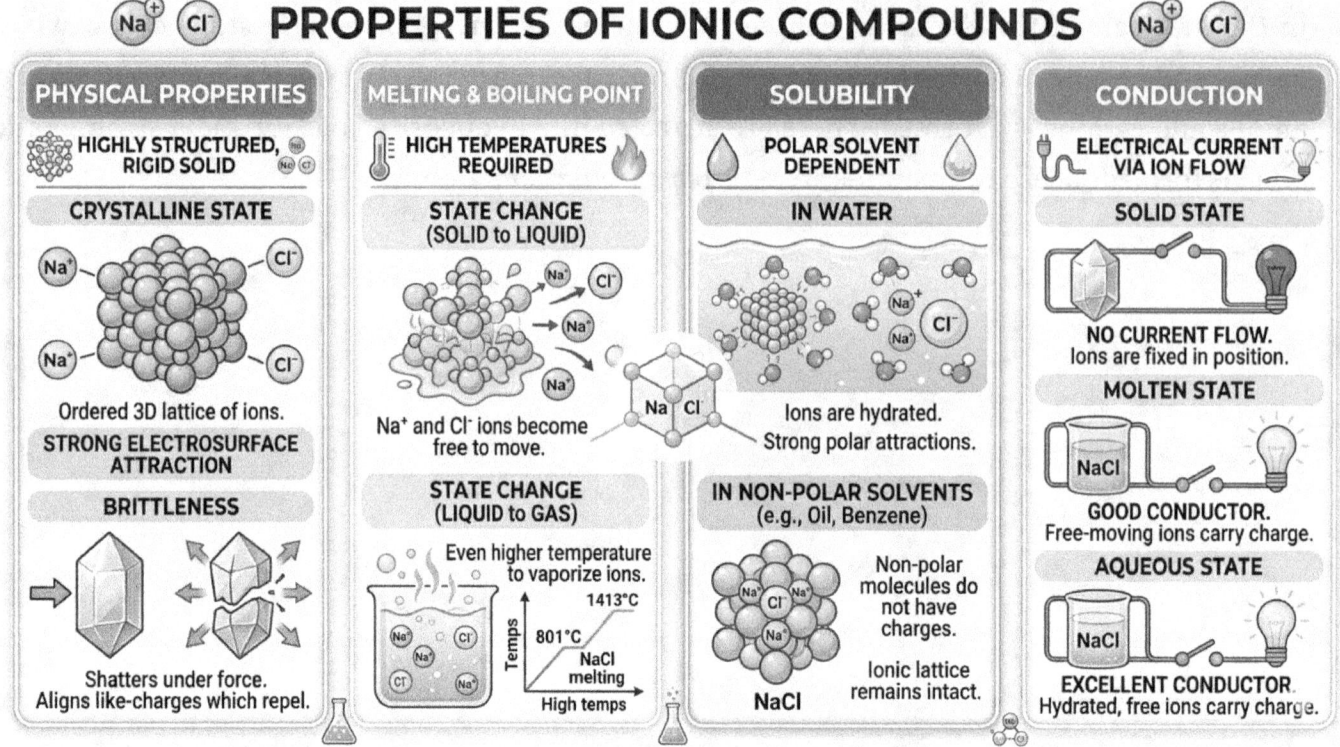

(i) **Physical nature**: Ionic compounds are solids and are somewhat hard because of the strong force of attraction between the positive and negative ions. These compounds are generally brittle and break into pieces when pressure is applied.

(ii) **Melting and Boiling points**: Ionic compounds have high melting and boiling points. This is because a considerable amount of energy is required to break the strong inter-ionic attraction.

(iii) **Solubility**: Electrovalent compounds are generally soluble in water and insoluble in solvents such as kerosene, petrol, etc.

(iv) **Conduction** *of Electricity*: The conduction of electricity through a solution involves the movement of charged particles. A solution of an ionic compound in water contains ions, which move to the opposite electrodes when electricity is passed through the solution. Ionic compounds in the solid state do not conduct

electricity because movement of ions in the solid is not possible due to their rigid structure. But ionic compounds conduct electricity in the molten state. This is possible in the molten state since the electrostatic forces of attraction between the oppositely charged ions are overcome due to the heat. Thus, the ions move freely and conduct electricity.

3.13 Ores and minerals

Compounds or elements which naturally occur in the earth's crust are known as **minerals**. The best example is water. Profitable minerals are referred as *ores*. If in getting a mineral we are in loss economically it's not an ore. The unwanted sand soil and mud contaminated with ore are referred as gangue.

3.14 Extraction of metals

Flow Chart: Extraction of Metals

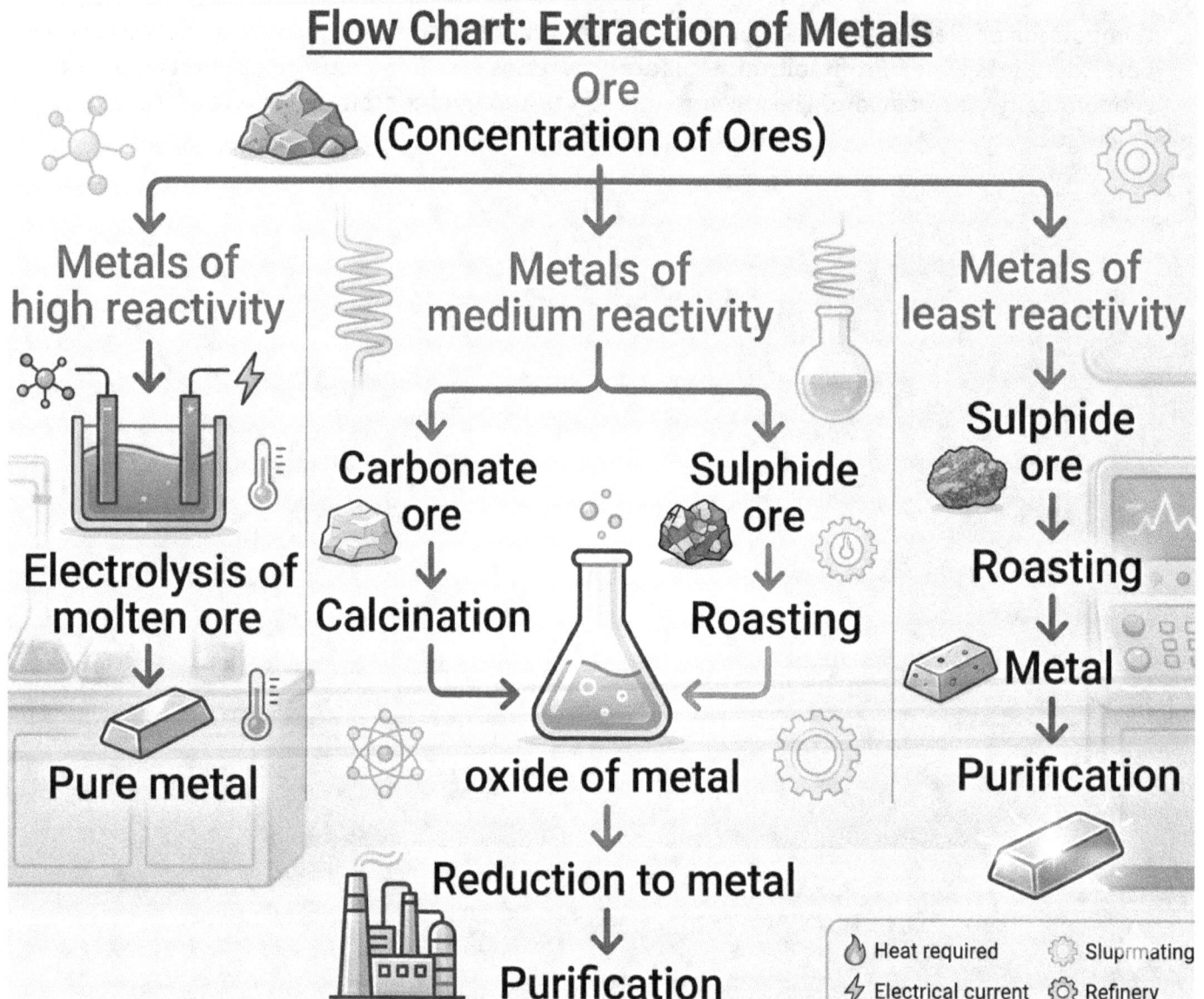

Ore
(Concentration of Ores)

Metals of high reactivity
→ Electrolysis of molten ore → Pure metal

Metals of medium reactivity
- Carbonate ore → Calcination
- Sulphide ore → Roasting
→ oxide of metal → Reduction to metal → Purification

Metals of least reactivity
Sulphide ore → Roasting → Metal → Purification

Heat required   Sluprmating
Electrical current   Refinery

3.15 Enrichment of Ores

Ores mined from the earth are usually contaminated with large amounts of impurities such as soil, sand, etc., called *gangue*. The impurities must be removed from the ore prior to the extraction of the metal. The processes several steps are involved in the extraction of pure metal from ores. A summary of these steps is given in flow chart. Each step is explained in detail in the following sections.

The processes used for removing the gangue from the ore are based on the differences between the physical or chemical properties of the gangue and the ore. Different separation techniques are accordingly employed.

3.16 Extracting Metals Low in the Reactivity Series

Metals lows in the activity series are very unreactive. The oxides of these metals can be reduced to metals by heating alone. For example, cinnabar (HgS) is an ore of mercury. When it is heated in air, it is first converted into mercuric oxide (HgO). Mercuric oxide is then reduced to mercury on further heating.

$$2HgS(s) + 3O (g) \rightarrow 2HgO(s) + 2SO_2 (g)$$
$$2HgO(s) \rightarrow 2Hg (l) + O_2 (g)$$

Similarly, copper which is found as Cu_2S in nature can be obtained from its ore by just heating in air.

$$2Cu S + 3O (g) \rightarrow 2Cu O(s) + 2SO (g)$$
$$2Cu O + 2Cu S \rightarrow 4Cu + 2SO2$$

3.17 Extracting Metals in the Middle of the Reactivity Series

The metals in the middle of the activity series such as iron, zinc, lead, copper, etc., are moderately reactive. These are usually present as sulphides or carbonates in nature. It is easier to obtain a metal from its oxide, as compared to its sulphides and carbonates. Therefore, prior to reduction, the metal sulphides and carbonates must be converted into metal oxides. The sulphide ores are converted into oxides by heating strongly in the presence of excess air. This process is known as *roasting*.

The carbonate ores are changed into oxides by heating strongly in limited air. This process is known as *calcination*. The chemical reaction that takes place during roasting and calcination of zinc ores can be shown as follows –

Roasting
$$2ZnS(s) + 3O(g) \rightarrow 2ZnO(s) + 2SO2(g)$$

Calcination
$$ZnCO3 \rightarrow (s) ZnO(s) + CO2(g)$$

The metal oxides are then reduced to the corresponding metals by using suitable reducing agents such as carbon.

For example,
when zinc oxide is heated with carbon, it is reduced to metallic zinc.

$$ZnO(s) + C(s) \rightarrow Zn(s) + CO(g)$$

Besides using carbon (coke) to reduce metal oxides to metals, sometimes displacement reactions can also be used. The highly reactive metals such as sodium, calcium, aluminium, etc., are used as reducing agents because they can displace metals of lower reactivity from their compounds. For example, when manganese dioxide is heated with aluminium powder, the following reaction takes place –

$$3MnO2(s) + 4Al(s) \rightarrow 3Mn(l) + 2Al2O3(s) + Heat$$

These displacement reactions are highly exothermic. The amount of heat evolved is so large that the metals are produced in the molten state. In fact, the reaction of iron (III) oxide ($Fe2O3$) with aluminum is used to join railway tracks or cracked machine parts. This reaction is known as the *thermit reaction*.

$$Fe_2O3(s) + 2Al(s) \rightarrow 2Fe(l) + Al_2O3(s) + Heat$$

3.18 Extracting Metals towards the
Top of the Reactivity Series

The metals high up in the reactivity series are very reactive. They cannot be obtained from their compounds by heating with carbon. For example, carbon cannot reduce the oxides of sodium, magnesium, calcium, aluminium, etc., to the respective metals. This is because these metals have more affinity for oxygen than carbon. These metals are obtained by electrolytic reduction.

For example,

sodium, magnesium and calcium are obtained by the electrolysis of their molten chlorides. The metals are deposited at the cathode (the negatively charged electrode), whereas, chlorine is liberated at the anode (the positively charged electrode). The reactions are –

At cathode $Na^+ + e- \rightarrow Na$ At

anode $2Cl^- \rightarrow Cl_2 + 2e^-$

Similarly, aluminium is obtained by the electrolytic reduction of aluminium oxide.

3.19 Refining of Metals

The metals produced by various reduction processes described above are not very pure. They contain impurities, which must be removed to obtain pure metals. The most widely used method for refining impure metals is electrolytic refining.

Electrolytic Refining: Many metals, such as copper, zinc, tin, nickel, silver, gold, etc., are refined electrolytically. In this process, the impure metal is made the anode and a thin strip of pure metal is made the cathode. A solution of the metal salt is used as an electrolyte. On passing the current through the electrolyte, the pure metal from the anode dissolves into the electrolyte. An equivalent amount of pure metal from the electrolyte is deposited on the cathode. The soluble impurities go into the solution, whereas, the insoluble impurities settle down at the bottom of the anode and are known as ***anode mud***.

Chapter 3: Deep Dive Challenges

Q1. Which metal is liquid at room temperature?

(a) Hg

(b) Br

(c) Cl

(d) None

Q2. Which nonmetal is liquid at room temperature?

(a) Hg

(b) Br

(c) Cl

(d) None

Q3. Which non metal is good conductor of electricity?

(a) Cu

(b) Fe

(c) Graphite

(d) Diamond

Q4. Which of the following can be cut by knife?

(a) Na

(b) Mg

(c) K

(d) All of the above

Q5. Calcination is used for?

(a) Sulphide ore

(b) Carbonate ore

(c) Oxide ore

(d) None of these

Q6. Which one is more reactive?

(a) Cu

(b) Pb

(c) Fe

(d) Zn

Q7. In thermite reaction the product is?

(a) Fe(l)

(b) Al_2O_3

(c) heat

(d) All of the above

Q8. In galvanization the layer of which metal is coated on iron or steel articles?

(a) Zn

(b) Mg

(c) K

(d) none of these

Q9. An alloy is a homogeneous mixture of?

(a) Two or more metals

(b) Metal and non-metals

(c) Both a and b

(d) None of these

Q10. Al_2O_3 is?

(a) Metallic oxide

(b) Nonmetallic oxides

(c) Amphoteric oxide

(d) None

Q11. Which types of substances are effective in cleaning the vessels?

(a) Sour

(b) Oily

(c) Sweets

(d) None of these

Q12. In electrolytic refining the pure metal is taken as?

(a) Anode

(b) Cathode

(c) Anyone can be taken

(d) None of these

Q13. Copper is used to make hot water tank because?

(a) It is good conductor of heat

(b) It is good conductor of electricity

(c) malleability

(d) Ductility

Q14. Coins are manufactured from silver and copper because?

(a) These are good conductor of heat

(b) These are good conductor of electricity

(c) Malleability

(d) Ductility

Q15. Nonmetallic oxides are?

(a) Acidic

(b) Neutral

(c) Either acidic or neutral

(d) None of these

Q16. Alloy of Cu and Zn is?

(a) Bronze

(b) Solder

(c) Brass

(d) Amalgam

Q17. In amalgam which element must be present?

(a) Hg

(b) Pb

(c) Fe

(d) Zn

Q18. solder has?

(a) High melting point

(b) Low melting point

(c) Moderate melting point

(d) Extremely low melting point

Q19. cinebar is Ore of?

(a) Hg

(b) Pb

(c) Fe

(d) None of these

Q20. Copper pyrite is ore of?

(a) Cu

(b) Pb

(c) Fe

(d) Zn

Match the following:

Q21.

A	B
Amphoteric	CO_2
Acidic	Al_2O_3
Basic	NaOH
Neutral	H_2O

Q22.

A	B
Ionic	CH_4
Co-valent	NaCl
Roasting	$ZnCO_3$
Calcination	ZnS

Q23.

A	B
Solder	Zn & Cu
Brass	Cu & Sn
Bronze	Hg & Al
Amalgam	Pb & Sn

Short Answer Questions

Q24. What are physical differences between metal and nonmetals?

Q25. What are chemical differences between metal and nonmetals?

Q26. What are alloys? Give examples.

Q27. What is difference between ores and minerals?

Q28. What are amphoteric oxides?

Q29. What is thermite reaction?

Q30. What happens when zinc reacts with cold water and steam?

Q31. Which type of compound is formed when metal reacts with nonmetals?

Q32. What is difference between calcinations and roasting?

Q33. How does metals are extracted from its oxides?

Long answer type questions

Q34. Give reasons

(a) Platinum, gold and silver are used to make jewellery.

(b) Sodium, potassium and lithium are stored under oil.

(c) Aluminium is a highly reactive metal, yet it is used to make utensils for cooking.

(d) Carbonate and sulphide ores are usually converted into oxides during the process of extraction.

Q35. Pratyush took sulphur powder on a spatula and heated it. He collected the gas evolved by inverting a test tube over it.

What will be the action of gas on:

(a) Dry litmus paper?

(b) Moist litmus paper?

(c) Write a balanced chemical equation for the reaction taking place.

Solution

1.a 2.b 3.c 4.d 5.b 6.d 7.d 8.a 9.c 10.c 11.a 12.b 13.a 14.c 15.c 16.c 17.a 18.b 19.a 20.a

---

Short Answer Hints

- Q24 (Physical): Think about "The 3 S's": Shininess (lustre), Sonority (ringing sound), and Strength (malleability/ductility). Also, consider which one is a better "highway" for heat and electricity.
- Q25 (Chemical): Focus on the oxides. When they react with oxygen, which one forms a basic oxide and which one forms an acidic one? Also, think about which loses electrons (+) and which gains them (-).
- Q26 (Alloys): It's a "mixture" of a metal with something else to make it stronger or rust-proof. Think of Steel, Brass, or the "Gold" used in jewelry.
- Q27 (Ores vs. Minerals): All ores are minerals, but not all minerals are ores. The key is profit—which one can you get the metal out of easily and cheaply?
- Q28 (Amphoteric): These are the "dual-natured" oxides. They can act like an acid when meeting a base, and like a base when meeting an acid. (Example: Aluminum or Zinc oxides).
- Q29 (Thermite): A very violent reaction used to join railway tracks. It involves Aluminum stealing oxygen from Iron oxide, releasing massive heat.
- Q30 (Zinc & Water): Zinc is "picky." It is too lazy to react with cold water. It needs more energy—does it react with hot water or only with the gas form (steam)?
- Q31 (Compounds): When a metal (giver) meets a non-metal (taker), they form a bond

based on attraction. Think of table salt (NaCl).

- Q32 (Calcination vs. Roasting): One is for Carbonate ores (think "C" for Calcination) and happens in limited air. The other is for Sulphide ores and needs "Roasting" in plenty of air.
- Q33 (Extraction): To get the metal, you must remove the Oxygen. This process is called Reduction. Think about using Carbon (Coke) or Electricity.

Long Answer Hints

Q34: Give Reasons

- (a) Jewelry: Why would you want a ring that doesn't tarnish or react with air and water even after many years?
- (b) Storage in Oil: These metals are "hyperactive." If they touch air or even a tiny drop of moisture, they might catch fire. Oil keeps them "sleeping."
- (c) Aluminum Utensils: Even though it's reactive, it quickly forms a protective invisible layer of oxide on its skin that stops further damage.
- (d) Conversion to Oxides: It is much easier to "pull" a metal out of an Oxide than it is to pull it out of a Carbonate or Sulphide.

Q35: The Sulphur Experiment

- The Gas: Heating Sulphur creates Sulphur Dioxide SO2).
- (a) Dry Litmus: Does a gas show its acidic nature without water to create ions?
- (b) Moist Litmus: When SO_2 meets the water on the paper, it forms Sulphurous Acid. What color do acids turn litmus?
- (c) Equation: $S + O_2 \rightarrow ?$ and then $SO_2 + H_2O \rightarrow ?$

NOTES

ABOUT THE AUTHOR

Umesh Kumar

Umesh Kumar is a professional educator, independent researcher, and author with extensive experience in academic counseling. A four-time UGC NET qualified professional in Electronic Science, he has served as a guest faculty member at the University of Delhi and as an empanelled counselor for IGNOU. Through his venture, Technic Point, he focuses on digital education and publishing, specializing in trilingual academic works and foundational study materials. Based in his native village of Nipaniya, Bihar, he is dedicated to simplifying complex concepts for students through his "Subtraction to Multiply" philosophy.

BOOKS BY THIS AUTHOR

High School Chemistry: The Complete Foundation

"Master the building blocks of the universe with this updated second edition by Assistant Professor Umesh Kumar. Spanning 9 comprehensive chapters, this book bridges the gap between the Indian CBSE curriculum and international scientific standards.

What's Inside:

Part 1: The Foundations – Deep dives into the Structure of the Atom, Laws of Chemical Equations, and the states of Matter.

Part 2: The Applications – Master Chemical Reactions, Acids & Bases, Metals, and the versatile nature of Carbon.

The Postal Address Method: A unique visualization tool for electronic configuration.

Modern Context: Insights into the chemistry of lithium-ion batteries, nanotechnology, and sustainable processes.

Competitive Edge: Refresh your knowledge with 'Deep Dive Challenges' at the end of every chapter."

www.ingramcontent.com/pod-product-compliance
Lightning Source LLC
Chambersburg PA
CBHW081303170526
45165CB00011B/3396